SHARKS Play Tricks!

BY JENNIFER LOMBARDO

ANIMAL TRICKSTERS

Gareth Stevens PUBLISHING

Please visit our website, www.garethstevens.com. For a free color catalog of all our high-quality books, call toll free 1-800-542-2595 or fax 1-877-542-2596.

Cataloging-in-Publication Data

Names: Lombardo, Jennifer.
Title: Sharks play tricks! / Jennifer Lombardo.
Description: New York : Gareth Stevens Publishing, 2025. | Series: Animal tricksters | Includes glossary and index.
Identifiers: ISBN 9781538293546 (pbk.) | ISBN 9781538293553 (library bound) | ISBN 9781538293560 (ebook)
Subjects: LCSH: Sharks–Juvenile literature. | Sharks–Behavior–Juvenile literature.
Classification: LCC QL638.9 L66 2025 | DDC 597.3–dc23

Published in 2025 by
Gareth Stevens Publishing
2544 Clinton Street
Buffalo, NY 14224

Designer: Andrea Davison-Bartolotta
Editor: Jennifer Lombardo

Photo credits: Cover, pp. 1, 7 Dirk van der Heide/Shutterstock.com; series art (background texture) Kriangkrai Namtongbai/Shutterstock.com; series art (shark skin texture) Vera Larina/Shutterstock.com; p. 5 BlueRingMedia/Shutterstock.com; p. 6 Valeri Luzina/Shutterstock.com; p. 9 Joe Dordo Brnobic/Shutterstock.com; p. 11 Illia Khurtin/Shutterstock.com; p. 13 (bottom) Blue Planet Archive/Alamy Stock Photo; p. 13 (top) courtesy of NOAA Photo Library/Flickr; p. 15 Jukka Jantunen/Shutterstock.com; p. 16 File:Pristiophorus cirratus.jpg/Wikimedia Commons; p. 17 (top) Auscape International Pty Ltd/Alamy Stock Photo; p. 17 (bottom) Pilschikova Ekaterina/Shutterstock.com; p. 18 3dsam79/Shutterstock.com; p. 19 (right) 9bdesign/Shutterstock.com; p. 19 (left) b-hide the scene/Shutterstock.com; p. 21 bearacreative/Shutterstock.com.

Printed in the United States of America

CPSIA compliance information: Batch #CSGS25: For further information contact Gareth Stevens at 1-800-542-2595.

CONTENTS

Words in the glossary appear in **bold** type the first time they are used in the text.

AMAZING SHARKS

Sharks are fish that have skeletons made of **cartilage** instead of bone. This makes them faster and lighter than other kinds of fish. We often think of sharks as being very big, but they come in all different sizes. The smallest is about the size of a candy bar!

There are about 500 species, or kinds, of sharks in the world. They have the same five senses humans do, plus two extra. They have special **organs** in their body that help them sense electrical currents and **pressure** changes in the water.

PARTS OF A SHARK

FUN FACT

A lateral line is a row of special cells that helps sharks sense pressure changes. Lateral lines help sharks find their way through dark water without bumping into things. They also help sharks find out where a smell is coming from.

Sharks come in all sizes, shapes, and colors, but most have the same basic parts.

THE SAND TIGER SHARK

Many fish have an organ in their body called a swim bladder. When they fill it with air, they float. When they let air out of it, they sink. Sharks don't have a swim bladder. If they stop swimming, they start to sink.

The sand tiger is the only species of shark that can swallow air. It can store the air in its stomach, making it act like a swim bladder. This lets sand tiger sharks float in the water without moving while they watch for **prey**!

Sand tigers look scary, but they're very calm around humans.

FUN FACT

In order to breathe, sharks always need water moving over their gills. Most sharks do this by swimming. When a sand tiger wants to float, it can use its throat to pump water over its gills.

THE SCALLOPED HAMMERHEAD SHARK

Most sharks are cold-blooded animals. This means that if the water they take in through their gills is cold, their body will get cold. If it's warm, they will warm up. The deeper the water is, the colder it gets.

The scalloped hammerhead preys on fish that live on the ocean floor, such as stingrays. It has a trick to stay warm while it dives for prey: It holds its breath! Scalloped hammerheads can close their gills to stop heat from leaving their body.

Scalloped hammerheads use their wide head to pin down prey so it can't escape.

THE SPOTTED WOBBEGONG SHARK

Some sharks swim around all day looking for food. However, spotted wobbegong sharks like to settle down and wait for the food to come to them! Their color helps **camouflage** them from both predators and prey. They have thin bits of skin around their upper lip that wave around and **lure** fish closer.

Wobbegongs have a very wide mouth, and they'll eat just about anything that can fit in it. If an animal is too large to swallow, the shark will chomp down with its teeth and hold the prey there until it gets too tired and dies.

Spotted wobbegongs are camouflaged to hide among rocks and coral reefs.

FUN FACT

The spotted wobbegong shark is also sometimes called the "carpet shark" because it lays flat on the ocean floor like a rug!

THE COOKIECUTTER SHARK

The cookiecutter shark also has a trick that lures prey to it. The largest of these sharks is only about 22 inches (56 cm) long, but they prey on animals more than 10 times their size! They eat sea creatures such as dolphins, stingrays, tuna, seals, and whales.

To lure such large, fast-moving prey, the cookicutter shark's body glows everywhere except for a band around its neck. To the shark's prey, this dark spot looks like a much smaller fish—prey for the prey. When the large animal gets close to try to eat the "fish," the shark attacks.

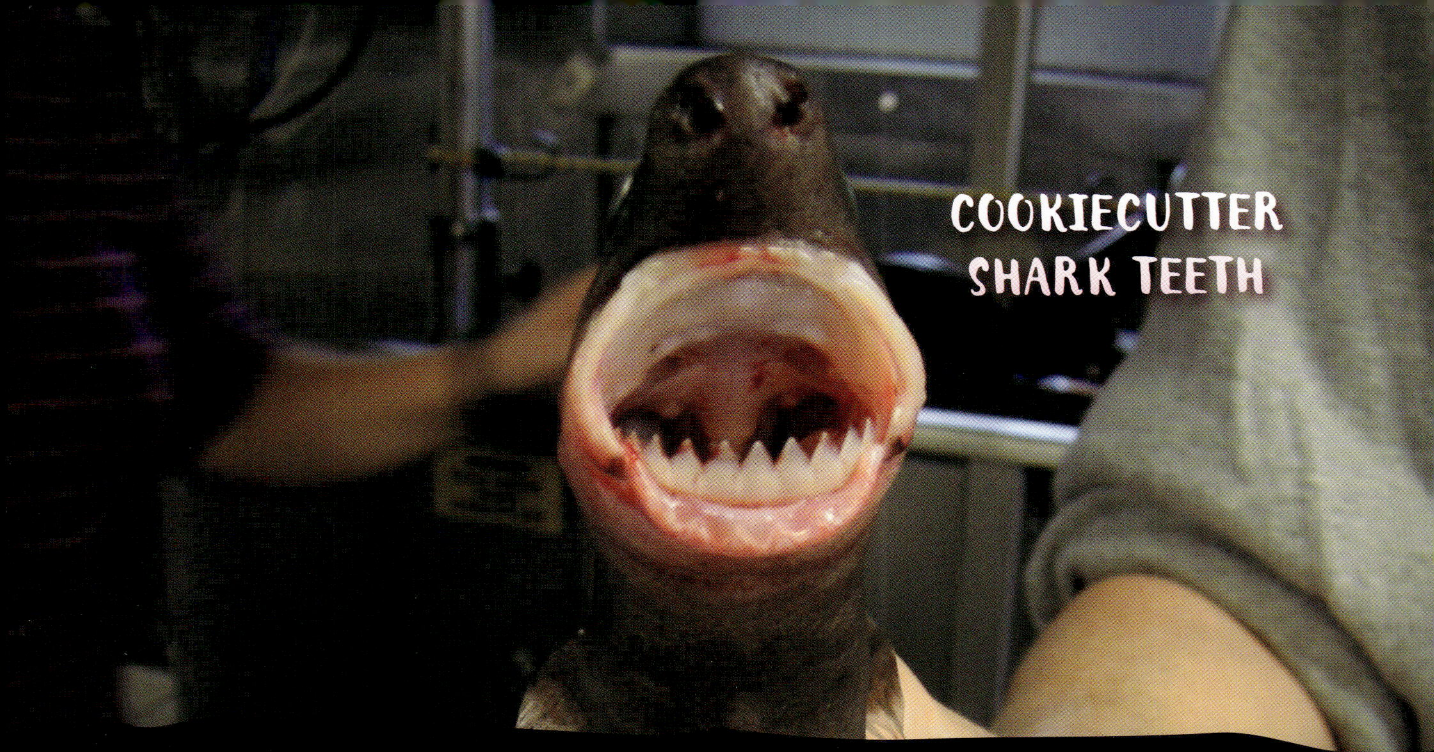

COOKIECUTTER SHARK TEETH

Scientists call the dark spot on a cookiecutter shark its "dog collar."

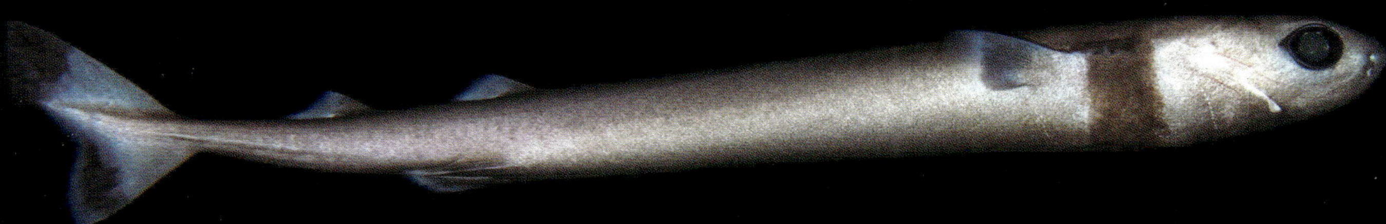

AN UNUSUAL ATTACK

Because the cookiecutter shark is so small, it doesn't eat most of its large prey whole. Instead, when the prey gets close, the shark attaches itself to its prey's body. It uses its lips and its upper teeth, which are small and pointy, to hang on. Then it spins its body around and uses its row of bottom teeth, which are much larger than the top teeth, to take a bite out of its prey.

The cookiecutter shark's unusual bite leaves a mark that is almost completely circular, like a cookie. This is how the shark got its name.

The circular scars on this sea lion's chest were made by a cookiecutter shark.

FUN FACT

One large animal the cookiecutter shark does sometimes eat whole is squid.

THE SAWSHARK

The sawshark gets its name from its long **snout**. Its length and the thin feelers on it make it look like a saw someone would use to cut wood. The sawshark has two tricks it can do with its snout. One is using it to fight off predators.

The other trick has to do with the sawfish's feelers. They can pick up electric signals that small fish give off. The sawfish runs its snout over the ground, feeling for those signals so it can flick its prey out of the sand and into its mouth.

The sawshark (top) looks a lot like the sawfish (bottom), but they're two different animals. The sawfish is actually a kind of ray!

THE GOBLIN SHARK

The goblin shark is one of the strangest-looking sharks! It has a long snout and sharp, crooked teeth. Goblin sharks live mostly at the bottom of the ocean off the coast of Japan.

The goblin shark doesn't swim very fast. To help it catch prey, it has an amazing trick. When prey is just out of reach, the goblin shark can shoot its jaw out of its mouth! The jaw is attached to bands of tissue called ligaments. When the shark is done eating, it fits its jaw back into its mouth.

The goblin shark got its name because when its jaw is outside its mouth, its snout looks like the long nose of a Japanese goblin.

THE THRESHER SHARK

The thresher shark is mainly found in the Atlantic Ocean. It's easy to recognize because of its long, oddly shaped tail. The thresher shark uses its tail to help it catch prey. When it finds a school, or group, of fish, the shark will swim around them. It makes smaller and smaller circles until the fish are very close together.

When the fish are in a tight group, the shark slaps them with its tail! This trick **stuns** them so the shark can catch them easily.

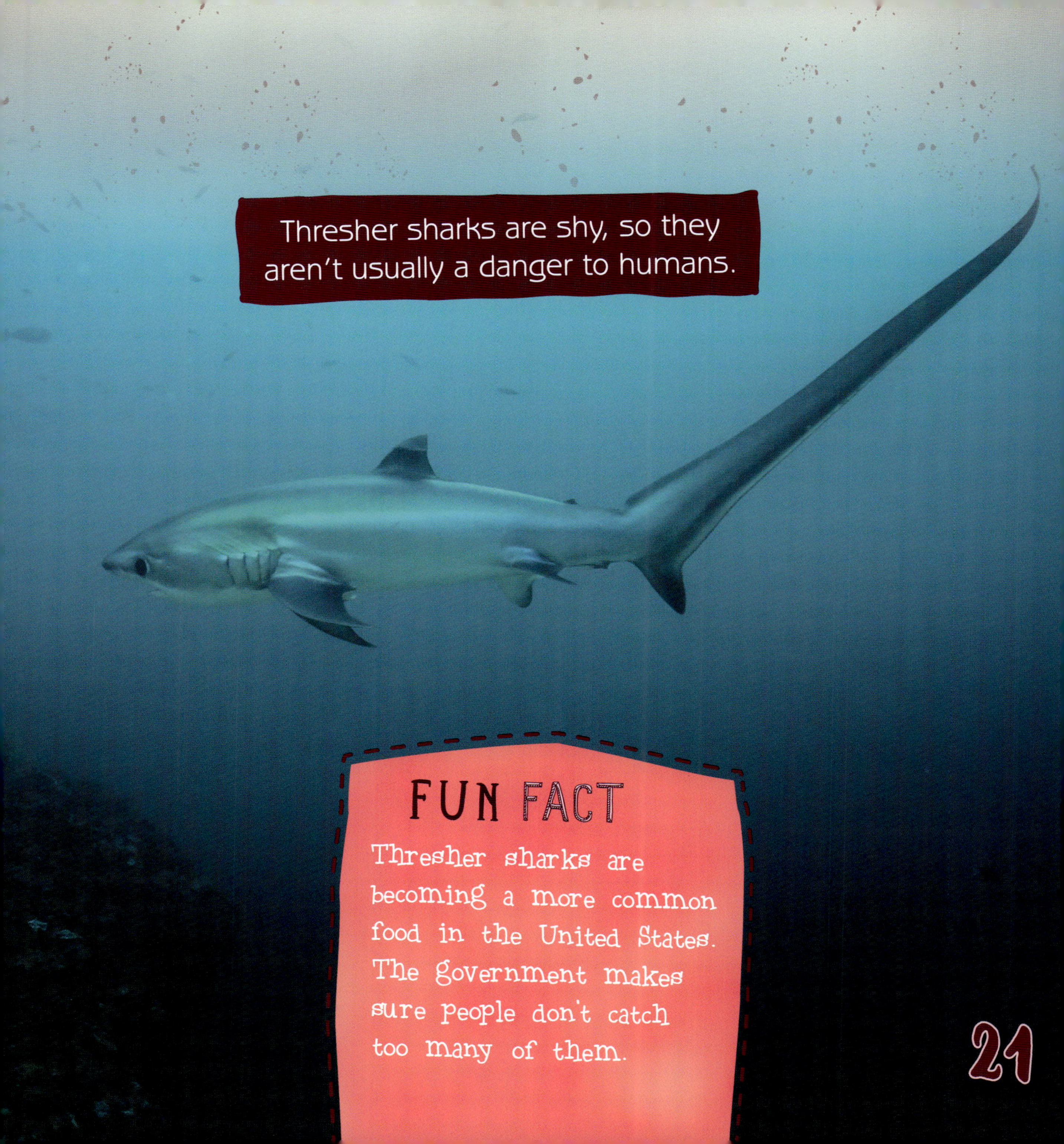

Thresher sharks are shy, so they aren't usually a danger to humans.

FUN FACT

Thresher sharks are becoming a more common food in the United States. The government makes sure people don't catch too many of them.

GLOSSARY

camouflage: Colors or shapes on animals that allow them to blend with their surroundings.

cartilage: Tough, bendable tissue that makes up part or all of a skeleton.

lure: To draw an animal closer in order to catch it.

organ: A part inside an animal's body.

pressure: A force that pushes on something else.

prey: An animal that is hunted by other animals for food.

snout: An animal's nose and mouth.

stun: To shock something so it can't move.

FOR MORE INFORMATION

Books

Gibbons, Gail. *Sharks.* New York, NY: Holiday House, 2020.

Leaf, Christina. *Sharks.* Minneapolis, MN: Bellwether Media, 2023.

Trimble, Irene. *Sharks.* Chicago, IL: Sequoia Kids Media, 2023.

Websites

Kiddle: Shark Facts for Kids
kids.kiddle.co/Shark
Learn more about sharks, and take a look at some awesome pictures.

KidZone: Sharks
www.kidzone.ws/sharks/facts.htm
This website gives some of the coolest shark facts.

National Geographic Kids: Goblin Shark
kids.nationalgeographic.com/animals/fish/facts/goblin-shark
Read more about the amazing goblin shark.

INDEX